SPECTACULAR WILD CATS

Written by Eliza Jeffery   Illustrated by Marina Halak

Copyright © 2024 Hungry Tomato Ltd

First published in 2024 by Hungry Tomato Ltd
F15, Old Bakery Studios, Blewetts Wharf, Malpas Road, Truro, Cornwall,
TR1 1QH, UK.

No part of this publication may be reproduced, stored in a retrieval system, or transmitted in any form or by any means, electronic, mechanical, photocopying, recording, or otherwise, without prior written permission of the copyright owner.

A CIP catalogue record for this book is available from the British Library.

ISBN 9781835693438

Printed in China

Discover more at
www.hungrytomato.com

CONTENTS

The World of Cats	4
Wild Cat Features	6
Spectacular Wild Cats	8
The Fastest	9
The Biggest	10
The Smallest	12
The Three Leopards	14
Weird and Wonderful	16
Habitat Cats	18
Wild Cat Facts	20
Extreme Cat Habitats	22
How To Protect Wild Cats	24
Name That Cat	26
What's That Cat?	28
Glossary	30
Index	31

Words in **BOLD** can be found in the glossary.

THE WORLD OF CATS

Get ready to explore the wonderful world of cats! From the big and powerful lion to the slim and speedy cheetah, there are so many different types of spectacular wild cats to discover.

WHAT IS A SPECIES?

A species is a group of living things, like animals or plants, that share **unique** characteristics. For example, lions and **domestic** cats are two different species. Around 40 species of cats exist, which can then be split into different breeds.

These are three different species!

Wild cats and domestic cats share a lot of features, but have very different personalities.

WHERE DO CATS COME FROM?

All cats are **descendants** of the African wildcat, a species believed to have emerged 12 million years ago! This cat is still around today, alongside many other types of wild cat. There are plenty of new species that have been domesticated by humans, too – these are the types of cats that we keep as pets!

LIVING IN THE WILD

Cats that live in the wild are the types of cat that have not been domesticated by humans, and can survive in their environments without our help. Many cats that live in the wild today, share characteristics of their wild cat **ancestors**. Over time, these wild cats have also **adapted** to have features that help them live in their changing **habitats**.

WILD CAT FEATURES

There are many features that wild cats have that make them stand out compared to their domesticated cat cousins. These special features allow them to survive in the wild.

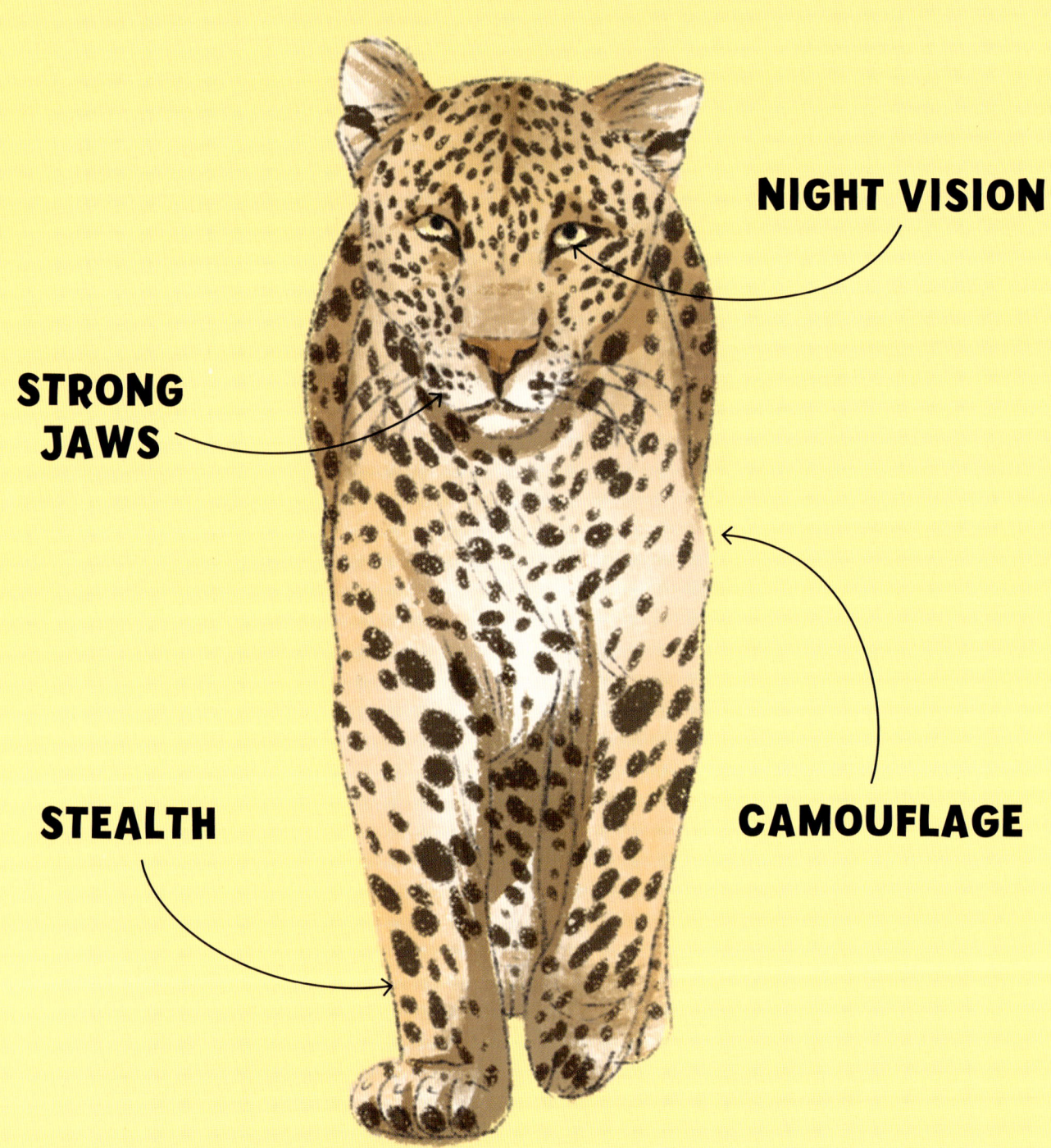

- **NIGHT VISION**
- **STRONG JAWS**
- **CAMOUFLAGE**
- **STEALTH**

NIGHT VISION

Many wild cats are **nocturnal**, meaning they wake up and hunt during the night. All cats have excellent night vision, but the ocelot is known for having the best night vision of all the wild cats.

STEALTH

Wild cats are known for being very **stealthy** animals! Big cats like leopards use slow, sneaky movements to hunt. They will often wait patiently in trees or by river edges before pouncing!

STRONG JAWS

Having strong jaws allows wild cats to easily catch and eat **prey**. Jaguars have the strongest jaws of all the wild cats, making their bite extra deadly! Powerful jaws make wild cats impressive hunters.

CAMOUFLAGE

Wild cats' coats help them blend into their natural habitats. Some, like tiger coats, have patterns, but others, like lion coats, are plain and match their surroundings. This allows them to travel without being seen by prey!

SPECTACULAR WILD CATS

From the little oncilla that spends its days pouncing on prey from trees to the powerful lion that lazes around by day and hunts at night, there are so many wild cats to explore. They live in lots of different habitats all around the world.

These types of cats live in the wild outdoors and do not rely on humans to survive. They are independent, deadly creatures that can't and shouldn't be tamed! It's time to delve into the world of spectacular wild cats...

Cheetah

The cheetah is not only the fastest cat, but the fastest animal in the world too! These speedy wild cats have golden fur and are covered in black spots. They might not pose a threat to humans, but you should stay far back when they're chasing prey!

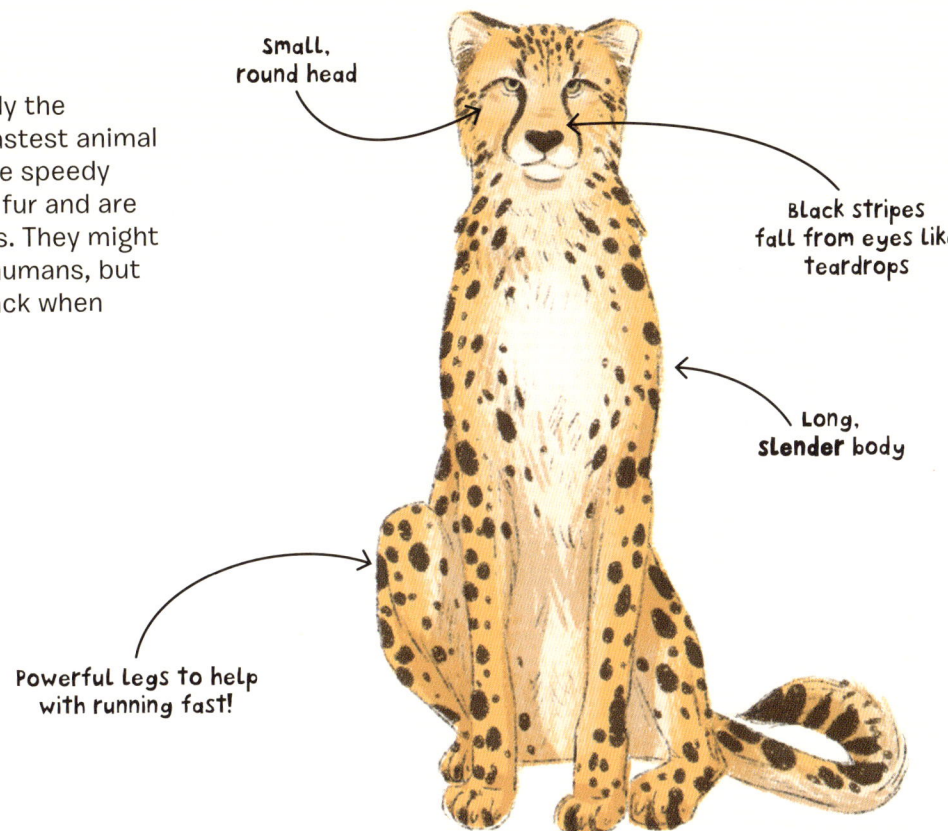

Small, round head
Black stripes fall from eyes like teardrops
Long, slender body
Powerful legs to help with running fast!

ORIGIN: Africa

COAT: Short and rough

PERSONALITY: Gentle and shy

WEIGHT	1 of 5 paws
SPEED	5 of 5 paws
ENDANGERED STATUS	3 of 5 paws

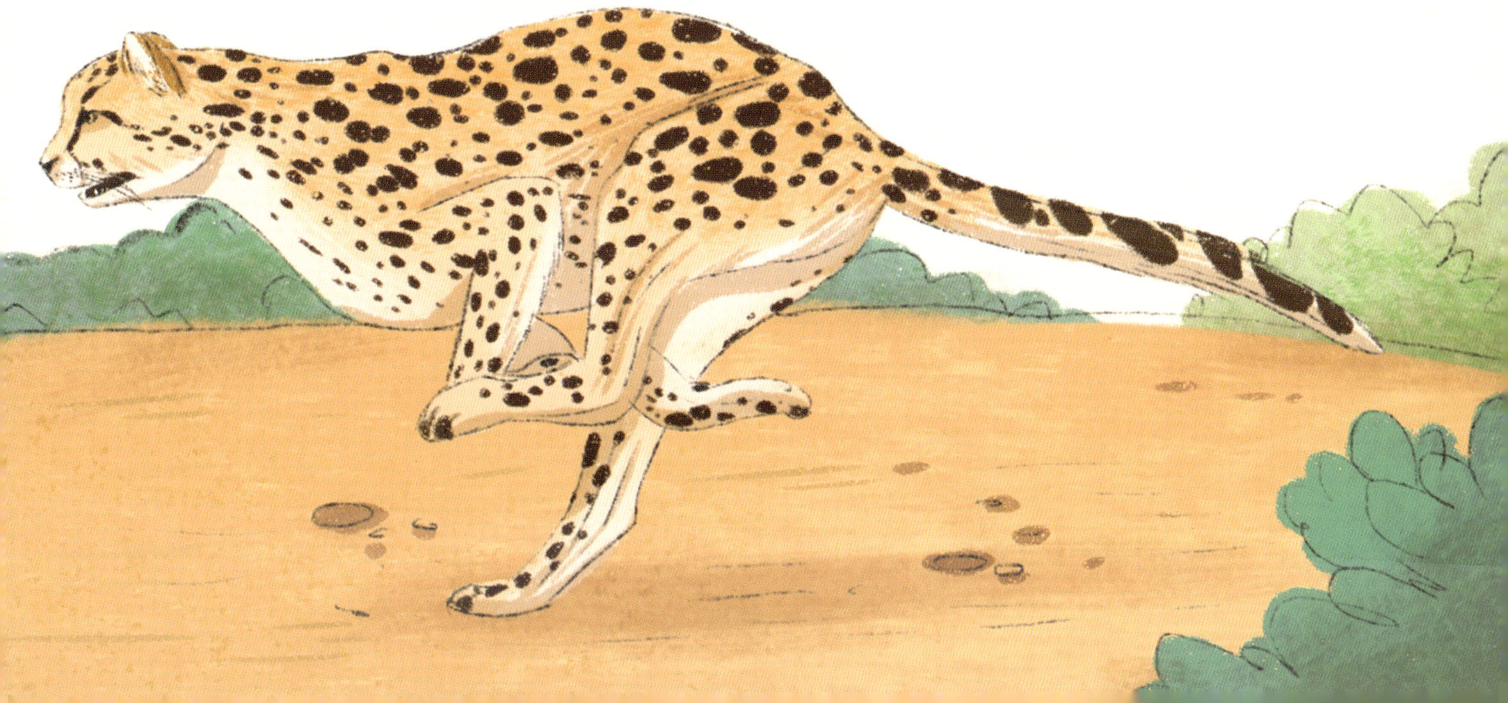

SPECTACULAR WILD CATS

Tiger

The tiger is the largest species of cat in the world! These strong, big cats, are easily identified by their orange coats covered in bold, black stripes. They tend to live alone and, unlike most cats, love swimming in water.

Spotty, round ears

Long whiskers

Large, padded feet

ORIGIN: Asia

COAT: Thick and coarse

PERSONALITY: Independent and fierce

WEIGHT 🐾🐾🐾🐾🐾

SPEED 🐾🐾🐾🐾🐾

ENDANGERED STATUS 🐾🐾🐾🐾🐾

Mountain Lion (Puma)

Mountain lions are known by many names, including puma, panther and cougar. These secretive wild cats are one of the largest carnivores in America, and are very hard to spot. They are very adaptable so can live in lots of different environments!

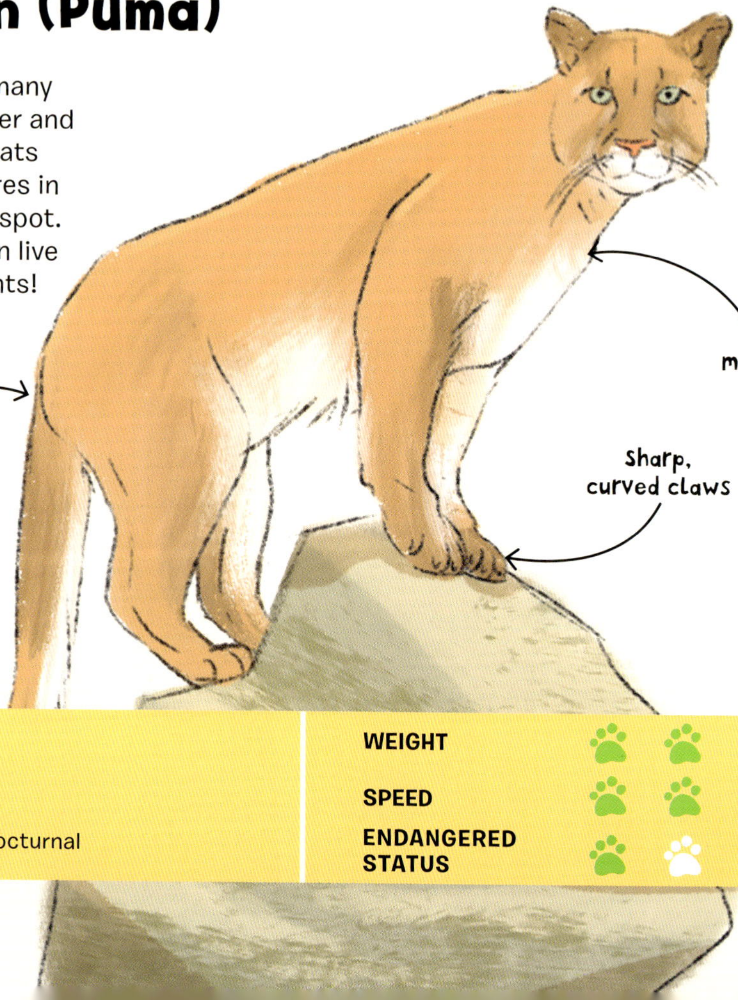

Light brown fur

Large, muscular body

Sharp, curved claws

ORIGIN: USA

COAT: Thick and soft

PERSONALITY: Solitary and nocturnal

WEIGHT 🐾🐾🐾🐾🐾

SPEED 🐾🐾🐾🐾🐾

ENDANGERED STATUS 🐾🐾🐾🐾🐾

THE BIGGEST   11

Lion

Lions are best known for their large manes and their powerful roar! They live in groups called prides and the males are nicknamed the "king of the jungle". These big cats may be deadly hunters but they often spend most of the day sleeping!

- Large, round head
- Only males have shaggy manes!
- Strong, muscular body

ORIGIN: Africa	**WEIGHT**
COAT: Short and coarse	**SPEED**
PERSONALITY: Powerful but lazy	**ENDANGERED STATUS**

Jaguar

Jaguars are impressive felines, with the ability to jump and climb up high in trees. They are also known for pouncing on their prey from way up high! These cats are excellent swimmers, and can often be found near the water's edge.

- Black spots all over its body!
- Short, sturdy legs
- Strong jaws, with a powerful bite!

ORIGIN: Africa	**WEIGHT**
COAT: Short and rough	**SPEED**
PERSONALITY: Independent and solitary	**ENDANGERED STATUS**

SPECTACULAR WILD CATS

Black-Footed Cat (Small-Spotted Cat)

The black-footed cat is considered the deadliest cat in the world. Despite being the smallest species of wild cat in Africa, this solitary cat searches for prey during the night and has a high success rate when it hunts!

Its eyesight is six times better than human's!

Under its coat, it has pink skin!

Dark spots on light fur

ORIGIN: South Africa

COAT: Soft and dense

PERSONALITY: Bold and independent

WEIGHT
SPEED
ENDANGERED STATUS

Rusty-Spotted Cat

Rusty-spotted cats are one of the smallest cat species in the world, only weighing up to 2 kg. Despite being small, this cat is known to hunt small **mammals** that are the same size as themselves! These tiny cats are very secretive, so little is known about how they behave.

Dark brown spots across body

Large, round eyes

Long, thick tail that is darker than the rest of the body

ORIGIN: India

COAT: Short and soft

PERSONALITY: Nocturnal and active

WEIGHT
SPEED
ENDANGERED STATUS

THE SMALLEST 13

Kodkod

Kodkods are the smallest wild cat in America and are closely related to the margay (page 17). They tend to live in rainforests and coastal forests. This shy feline spends lots of its time awake, resting in the day and then exploring and hunting at night.

They are usually brown and spotty, but can have black coats too!

Long, slender body

Black spots all over fur, even if you can't see them!

ORIGIN: Chile	**WEIGHT** 🐾
COAT: Thick and dense	**SPEED** 🐾
PERSONALITY: Secretive and solitary	**ENDANGERED STATUS** 🐾🐾🐾

Oncilla

Oncilla are one of the smallest wild cats in South America, hunting at night and choosing to live a solitary life. This feisty wild cat will hide in trees and look for its prey down below, waiting for the perfect opportunity to pounce!

Eyes vary from light to dark brown

Dark brown spots, outlined by black circles

Fantastic climber!

ORIGIN: Central and South America	**WEIGHT** 🐾
COAT: Short and thick	**SPEED** 🐾🐾🐾
PERSONALITY: Aggressive and independent	**ENDANGERED STATUS** 🐾🐾

SPECTACULAR WILD CATS

Leopard

Similar to a jaguar (page 11) in appearance, the leopard is a skilled hunter that only comes out at night. These powerful felines are great climbers, and are able to pull their prey high up into the trees! They eat all sorts of prey, from little bugs to big antelopes.

Powerful, loud roar!

Sturdy legs for jumping

Large paws with sharp claws

ORIGIN: Africa
COAT: Thick and soft
PERSONALITY: Solitary and vocal

WEIGHT	🐾 🐾 🐾 🐾 🐾 (1/5)
SPEED	🐾 🐾 🐾 🐾 🐾 (3/5)
ENDANGERED STATUS	🐾 🐾 🐾 🐾 🐾 (2/5)

Leopards like to lie in the sun!

Despite their names, snow leopards and clouded leopards aren't leopards - they're different species entirely!

THE THREE LEOPARDS 15

Clouded Leopard

The clouded leopard is one of the oldest species of cat in the world. This wild cat is a very clever and quiet **predator**, and can't roar or purr. They are very rarely seen by humans, so little is known about this secretive wild cat.

Has special pads on its feet to help grip to branches

Long, slender body

Extremely long tail!

ORIGIN: Asia	**WEIGHT** 🐾🐾🐾
COAT: Soft and dense	**SPEED** 🐾🐾🐾🐾
PERSONALITY: Nocturnal and solitary	**ENDANGERED STATUS** 🐾🐾🐾🐾

Snow Leopard

Snow leopards are experts at **camouflage**! They have beautiful thick fur that blends in with their snowy habitat, keeping them hidden from unsuspecting prey. They have large paws covered in thick fur that keep them warm against the snow.

Well adapted to the cold weather!

Long, bushy tail

Strong, powerful legs for jumping

ORIGIN: Asia	**WEIGHT** 🐾
COAT: Thick and soft	**SPEED** 🐾🐾🐾
PERSONALITY: Shy and solitary	**ENDANGERED STATUS** 🐾🐾🐾🐾

SPECTACULAR WILD CATS

Serval

Servals have very large, tall ears and look like a small version of a cheetah (page 9)! They are active cats that jump high to catch birds above and dig deep into the ground to find prey below. Despite their small size, these clever felines can reach speeds of 45 mph (72 km/h).

Long neck

Small head

strong, slender legs

ORIGIN: Africa	**WEIGHT** 🐾 🐾 🐾 🐾 🐾
COAT: Soft and dense	**SPEED** 🐾 🐾 🐾 🐾 🐾
PERSONALITY: Independent and intelligent	**ENDANGERED STATUS** 🐾 🐾 🐾 🐾 🐾

Caracal

Caracals are easy to identify by the tall tufts of fur at the top of their ears! These are believed to help them blend into the tall grass that covers their habitat, as well as help them to communicate with each other – they do this by twitching their tufts!

Powerful back legs for jumping

Slender, sturdy body

Likes to catch birds!

ORIGIN: Africa	**WEIGHT** 🐾 🐾 🐾 🐾 🐾
COAT: Short and soft	**SPEED** 🐾 🐾 🐾 🐾 🐾
PERSONALITY: Secretive and intelligent	**ENDANGERED STATUS** 🐾 🐾 🐾 🐾 🐾

WEIRD AND WONDERFUL

Margay

Margay have unique abilities – they can rotate their ankles 180 degrees. This makes them excellent climbers, and means these flexible felines can climb down trees headfirst. The margay is the monkey of the cat world!

- can often be found high up in the trees!
- Distinctive, black markings on face
- Large, brown eyes

ORIGIN: Central and South America

COAT: Thick and dense

PERSONALITY: Nocturnal and active

WEIGHT 🐾
SPEED 🐾
ENDANGERED STATUS 🐾🐾

Pallas's Cat

The Pallas's cat is an unusual-looking feline, easy to spot by its flattened, round face, sweet expressions and massive coat of fur. They have the longest coat of any species of cat. Pallas's cats spend most of their time in caves, burrows or around rocky areas.

- Ears sit flat against their head!
- Faint stripes across body
- Short, strong legs

ORIGIN: Asia

COAT: Soft and thick

PERSONALITY: Shy and solitary

WEIGHT 🐾
SPEED 🐾
ENDANGERED STATUS 🐾

SPECTACULAR WILD CATS

Lynx

Lynx are small wild cats that live in thick woodlands and rocky mountains. The tufts on their ears are believed to help them hear better, and their fur thickens in the winter to keep them warm. They are independent cats, perfectly built for the cold weather!

Black tufts on the top of the ears!

Lightly spotted fur

Large, **webbed** paws to stop them from slipping

ORIGIN: Europe and Africa

COAT: Soft and thick

PERSONALITY: Reserved and solitary

WEIGHT 🐾🐾⬜⬜⬜

SPEED 🐾🐾🐾🐾⬜

ENDANGERED STATUS 🐾🐾⬜⬜⬜

Bobcat

The bobcat is best known for its short bobble tail! These adaptable felines are capable of living in a wide variety of habitats; from forests and deserts, to coastal areas and scrubland, there is nowhere these cats won't go!

Small, pointy ears

Distinctive markings on face

Black stripes or spots on legs

ORIGIN: North America

COAT: Short and dense

PERSONALITY: Independent and fierce

WEIGHT 🐾⬜⬜⬜⬜

SPEED 🐾🐾⬜⬜⬜

ENDANGERED STATUS 🐾⬜⬜⬜⬜

HABITAT CATS   19

Flat-Headed Cat

As the name suggests, this striking feline is easy to distinguish by its flat head, as well as its constantly shocked expression! This cat is never too far away from water, choosing to live by marshes, streams, lakes, wetlands, and in rainforests.

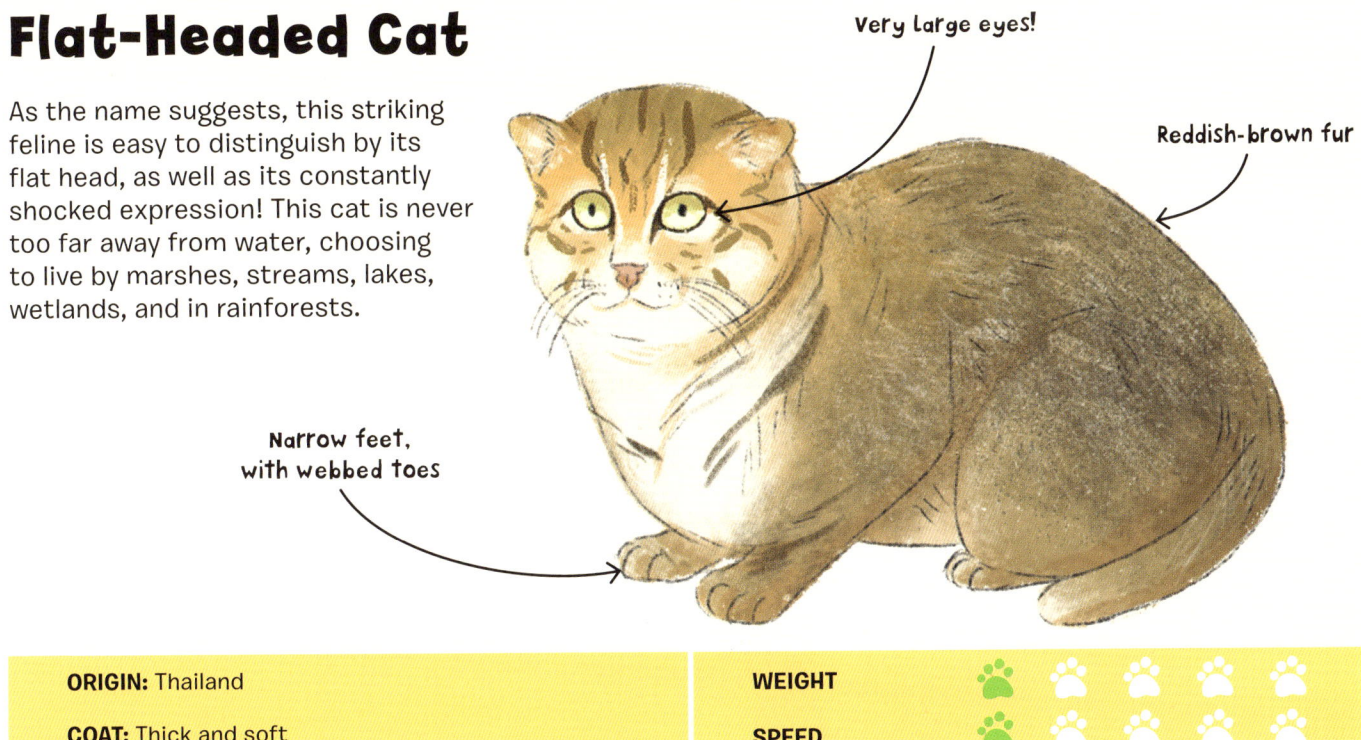

Very large eyes!

Reddish-brown fur

Narrow feet, with webbed toes

ORIGIN: Thailand	**WEIGHT** 🐾 🐾
COAT: Thick and soft	**SPEED** 🐾 🐾
PERSONALITY: Nocturnal and shy	**ENDANGERED STATUS** 🐾 🐾 🐾

Sand Cat

The sand cat is well adapted to sweltering climates, living in deserts across Africa and Asia. Their pale fur allows these cute felines to blend in with their environment. These impressive cats can go weeks without drinking any water!

Dark fur markings, from eyes across face

Brown stripes on legs and tail

Sharp claws

ORIGIN: Africa	**WEIGHT** 🐾
COAT: Soft and dense	**SPEED** 🐾 🐾
PERSONALITY: Solitary and vocal	**ENDANGERED STATUS** 🐾

WILD CAT FACTS

What else is there to know about the incredible world of wild cats? Let's uncover more intriguing facts about these majestic and powerful creatures.

SMALL BUT DEADLY

Despite being one of the smallest cats living in the wild, the black-footed cat is a more skilful hunter than a lion! It can jump 1.5 metres in the air and eats about 3,000 rodents a year!

They may look cute, but they are very scary if you are a small animal!

MOUNTAIN LION SCREAM

These solitary felines don't roar like lions and tigers do; they scream! Female mountain lions in particular will let out a high-pitched scream to attract a male.

SNOW LEOPARD SCARF

Snow leopards' long, fluffy tails helps them balance in thick snow. They also use their tails like scarves, wrapping them around themselves for extra warmth when it's cold.

Snow leopards have very thick, bushy tails.

CAN YOU SMELL THAT?

Cats have an excellent sense of smell. Their sense of smell is 14 times better than humans'! Tigers are known for having the best sense of smell of all the cats. They use smell to mark their **territory**, and to know when prey is nearby!

CAT OR MONKEY?

Cats can develop many special skills to help them hunt for food. For example, some felines attract their prey by copying the sounds they make. Margays are known for making monkey noises to catch their attention, before pouncing!

EXTREME CAT HABITATS

Cats are incredibly adaptable creatures that can live in all sorts of habitats. Wild cats can even live in some of the most extreme environments around the world!

DESERT HOME

Sand cats have adapted to thrive in extreme desert heat. Their large, tufted ears help them to hear but, also help them to cool down. Sand cats get most of their water from the animals they eat, like lizards and mice, which helps these clever cats live in the dry desert.

They have special fur that helps them stay cool during the day, and warm during sandstorms.

MOUNTAIN HOME

Snow leopards live in mountain habitats. They have short front legs and longer back legs that help them to climb in the challenging environments they hunt in.

Despite their name, they are more closely related to the tiger!

RAINFOREST HOME

Oncillas are experts at surviving in their rainforest home. They are excellent climbers and often use the tall trees for hunting, resting, and staying safe from bigger predators.

Their spotty fur helps them blend in with the forest floor.

SHRUBLAND HOME

The Pallas's cat thrive in their rocky, shrubland home. They build dens in natural rock crevices and feed off the prey that lives there. They also don't need to take in too much water.

They have short legs and are better at stalking than running!

HOW TO PROTECT WILD CATS

There are many species of cat that live in the wild that are endangered or under threat. It's super important that we do things to protect these animals so they can stay healthy and safe for a long, long time.

KEEP HABITATS SAFE

Big cats like tigers, lions, and cheetahs need lots of space to roam and hunt. Governments can help protect their grassland and forest homes, by preventing roads being built and trees being cut down - things which would harm their natural habitat. Making these small changes could have a big impact on the homes of wild cats.

STOP HUNTING

We need to make sure people don't harm them. One way we can protect them is by telling people that it's not fair to hurt these amazing animals and to stop hunting them. Hunting these cats makes them endangered!

Big cats like tigers or lions are hunted for their fur or because people think they are dangerous!

BE KIND TO NATURE

By looking after our world's natural habitats and all the animals in it, we can keep the balance in the wild. This makes sure that there's enough food and homes for everyone, including wild cats!

Looking after our planet will help more than just wild cats!

EDUCATE THE WORLD

We can help save wild cats through educating our communities about ways to live alongside them safely! Learning about wild cats is important because it helps us to understand these mysterious animals better. The more we know about them, the more we can help them! If we all take the right steps to learn about these felines, we can make a difference to their future.

NAME THAT CAT

Can you work out which wild cat each of these pictures are a part of? Clues have been provided for you based on facts in this book.

CLUE: These wild cats can climb down trees head first.

CLUE: These wild cats are the largest of them all.

CLUE: These wild cats have tall tufts of fur at the top of their ears to help them blend into the tall grass.

CLUE: These wild cats have very large, tall ears, and look like smaller versions of cheetahs.

CLUE: These wild cats are considered the deadliest cats in the world.

CLUE: These male wild cats have shaggy manes.

CLUE: These wild cats are known to hunt prey the same size as them.

CLUE: These wild cats are the fastest animal in the world.

CLUE: These wild cats are well-known for their flat heads and shocked expressions.

CLUE: These wild cats are experts at camouflaging in their snowy habitat.

Answers can be found on page 32.

WHAT'S THAT CAT?

Now that you have read all about these spectacular cats, how good are you at identifying them? There are 20 different cats to figure out. Use the information in the book to help you.

What am I?
A. Cheetah
B. Pallas's Cat
C. Oncilla

What am I?
A. Pallas's Cat
B. Singapura
C. Mountain Lion

What am I?
A. Flat-headed Cat
B. Leopard
C. Lion

What am I?
A. Sand Cat
B. Cheetah
C. Mountain Lion

What am I?
A. Margay
B. Mountain Lion
C. Serval

What am I?
A. Kodkod
B. Lynx
C. Oncilla

What am I?
A. Leopard
B. Sand Cat
C. Serval

What am I?
A. Jaguar
B. Serval
C. Snow Leopard

What am I?
A. Caracal
B. Tiger
C. Clouded Leopard

What am I?
A. Pallas's Cat
B. Lion
C. Black-footed Cat

Answers can be found on page 32.

What am I?
A. Lynx
B. Caracal
C. Leopard

What am I?
A. Tiger
B. Rusty-Spotted Cat
C. Margay

What am I?
A. Lynx
B. Caracal
C. Snow Leopard

What am I?
A. Lynx
B. Mountain Lion
C. Jaguar

What am I?
A. Rusty-spotted cat
B. Snow Leopard
C. Kodkod

What am I?
A. Tiger
B. Bobcat
C. Jaguar

What am I?
A. Mountain Lion
B. Flat-headed cat
C. Sand Cat

What am I?
A. Margay
B. Black-footed Cat
C. Lion

What am I?
A. Flat-headed Cat
B. Jaguar
C. Bobcat

What am I?
A. Cheetah
B. Clouded Leopard
C. Kodkod

GLOSSARY

Adapted - being able to adjust to new surroundings.

Ancestors - the people who came before us in our family tree.

Camouflage - when animals are able to blend in with their surroundings.

Descendants – people or animals that are related to an individual or group who lived in the past. For example, you are a descendant of your parents and grandparents.

Domestic – an animal that has been tamed or trained to live or work with humans.

Endangered - when a species of any kind is at risk of no longer existing.

Environment - another word for surroundings.

Habitats - the natural environment where animals, plants, and any other things live.

Mammals - warm-blooded animals that give birth to live young and feed their babies with milk.

Nocturnal - when an animal sleeps during the day and becomes active during the night.

Predator - an animal that hunts and kills other animals for food.

Prey - animals that are hunted and killed for food.

Slender - something that is thin and narrow.

Solitary - something that lives alone.

Stealthy - when something moves carefully and quietly, like a tiger hunting its prey without making a sound.

Territory - a specific area that belongs to someone or something.

Unique - something that stands out and is completely different from everything else.

Webbed - toes that are connected by a thin piece of skin. Some animals have these to help them swim.

INDEX

B
Black-footed cat (Small-spotted cat) 12, 20, 28-29, 32
Bobcat 18, 28-29, 32

C
Caracal 16, 28-29, 32
Cheetah 4, 9, 16, 24, 26, 28-29, 32
Clouded leopard 14-15, 28-29, 32

E
extreme habitats 22-23

F
features 5, 6-7
    camouflage 6-7, 15, 27, 30
    night vision 6-7
    stealth 6-7, 30
    strong jaws 6-7, 11
Flat-headed cat 19, 28-29, 32

H
hunting 7, 11, 12-13, 14, 20-21, 23, 24

J
Jaguar 6-7, 11, 14, 28-29, 32

K
Kodkod 13, 28-29, 32

L
Leopard 7, 14, 28-29, 32
Lion 4, 7, 8, 11, 20, 24, 28-29, 32
Lynx 18, 28-29, 32

M
Margay 13, 17, 28-29, 32
Mountain lion (puma) 10, 20, 28-29, 32

O
Oncilla 8, 13, 23, 28-29, 32

P
Pallas's cat 17, 23, 28-29, 32

R
Rusty-spotted cat 12, 28-29, 32

S
Sand cat 19, 22, 28-29, 32
Serval 16, 28-29, 32
Snow leopard 14-15, 21, 22, 28-29, 32
species 4-5, 10, 12, 14-15, 17, 24, 30

T
Tiger 7, 10, 20-21, 22, 24, 28-29, 30, 32

NAME THAT CAT ANSWERS

1 - Margay
2 - Tiger
3 - Caracal
4 - Serval
5 - Black-footed Cat
6 - Lion
7 - Rusty-spotted Cat
8 - Cheetah
9 - Flat-headed Cat
10 - Snow Leopard

WHAT'S THAT CAT ANSWERS

1 - A. Cheetah
2 - A. Pallas's Cat
3 - B. Leopard
4 - C. Mountain Lion
5 - A. Margay
6 - C. Oncilla
7 - B. Sand Cat
8 - B. Serval
9 - C. Clouded Leopard
10 - C. Black-footed Cat
11 - B. Caracal
12 - A. Tiger
13 - C. Snow Leopard
14 - A. Lynx
15 - A. Rusty-spotted Cat
16 - B. Bobcat
17 - B. Flat-headed Cat
18 - C. Lion
19 - B. Jaguar
20 - C. Kodkod

Picture Credits:
(abbreviations: t=top, b=bottom, m=middle, l=left, r=right)

Wikipedia: Mauro Tammone 29br. Shutterstock: Alexandr Junek Imaging 28mr; Anil Varma 29tl; B.Allen 28bl; Bharat Goel Photographer 21mr; Daniel J.Rao 23br; Dennis W Donohue 29tr; Dirk Brink 20br; Eumates 29tl; Felineus 29ml; FOTGRIN 21b; Ground Picture 25bl; Heinrich Neumeyer 7tr; Josef_Svoboda 29ml, 28ml; Jurgens Potgieter 24tr; Kyslynskahal 28tr; Maria Bukval 22br; Mark Green 29bl; Los t 7bl; PaniYani 28tl; Risto Raunio 28tr; Ruek66 28mr; Saad315 7tl; Signature Message 29mr; Slowmotiongli 20tl, 28br, 28ml; Stanlet Dullea 22tl; Sunshine Seeds 7br; Teo Tarras 23tl; TYRERPIX 29mr; VesnaArt 25tr; WildMedia 29tr; Wildsight 28tl.

Every effort has been made to trace the copyright holders, and we apologise in advance for any unintentional omissions. We would be pleased to insert the appropriate acknowledgements in any subsequent edition of this publication.

ABOUT THE AUTHOR

Eliza Jeffery is a children's book author based in Falmouth. She is passionate about helping children explore and enjoy the big world around them. She loves exploring Cornwall, and can often be found reading a book and eating a bowl of mussels by the sea!

ABOUT THE ILLUSTRATOR

Marina Halak is a talented illustrator of children's books from Ukraine. Her stunning illustrations are inspired by her own childhood, children, nature, magical moments and fairy tales. Marina is also the illustrator behind the series, *Dogs*.